Fort Victoria is
Westhill Lane,
OS Map: SZ

Leave the car park via the arch next to the Planetarium. Turn right along the sea wall and walk along the shingle bed to join the concrete sea wall. Continue along the sea wall to the end, to a wooden defence and the tidal salt marsh. Turn right (signposted "Coastal Path") uphill to the main road. Cross the main road and turn left along the pavement and cross the bridge. Turn right immediately over the bridge on to a tarmac path and continue ahead around the estuary to join the gravel track leading towards the old mill. Continue ahead along the sea wall, past the old mill, and pass through a swing gate onto a gravel track. Turn right (signposted "Y19") and continue ahead to the road at the Causeway, which has lovely views down the river Yar towards Yarmouth.

The Causeway with All Saints' church.

In Hooke's day, the West Wight was a separate island, the sea defences and the narrow neck of land bearing the road past Freshwater Bay is all that prevents it being so now. As a boy, Robert Hooke developed an aptitude for building mechanical devices by imitating local craftsmen. In his notebook he describes how "...*he made a small ship about a Yard long, fitly shaping it, adding*

Rigging of Ropes, Pullies, Masts, &c. with a contrivance to make it fire off small Guns, as it was sailing cross a Haven of pretty breadth". The Haven was on the sluggish river Yar, close to here.

▶ **Turn right over the Causeway and continue ahead uphill to All Saints' Church.**

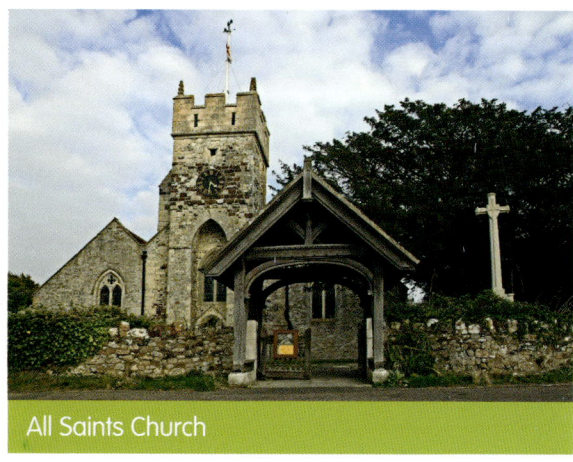

All Saints Church

Robert Hooke was born on 18th July 1635 in Freshwater, the son of John Hooke, who was the curate of the Parish church of All Saints', Freshwater, and John's second wife, Cecelie (nee Gyles),who came from Brading. The Church is one of the oldest established churches on the Island. There is evidence that nine hundred years ago there was a simple two-cell Anglo-Saxon church on this site. Part of this original structure can be seen near the pulpit where pillars are supported on quoins, typical of the period. In the 12th and 13th century the church was extended as the area became wealthier. The church underwent major restoration and expansion in 1875-6 during which time the plaster was removed to expose the stonework.

Robert Hooke was a sickly infant, who for seven years was not expected to survive, but he gradually grew stronger, becoming agile and energetic, though not robust. He was a fast learner, so his father hoped that he might make a career in the church but Robert's weak constitution and his father's increasing ill-health ended this ambition, and the boy was left to his own devices. Robert followed his own inclinations, with the result that the Island formed him as a scientist.

In the 1640's John Hoskins, a famous painter of miniatures, visited the Island. Hooke's friend, John Aubrey, recounts how, *"Mr Hooke observed what he did...So he gets him chalke, and ruddle (red ochre), and coale, and grinds them, and puts them on a trencher, gott a pencil, and to worke he went, and made a picture: then he copied (as they hung up in the parlour) the pictures there."* The skill he acquired with pen and pencil was invaluable later on when he needed to communicate his microcopical and mechanical discoveries, and in his career as an architect and surveyor.

All Saints' Church was the centre of a thriving village located here because of the river crossing further down the lane.

▶ **Continue ahead along the road to the T-junction of Copse Lane and Hooke Hill.**

Hooke's Childhood Home

The family lived in a modestly furnished cottage with a parlour, hall, study, kitchen and buttery downstairs and three attic bedrooms above them. This cottage, which no longer exists, was on Hooke Hill. It was situated behind what is now Heatherstone House. Robert would have shared one of the bedrooms with his brother John, until the older boy became a grocer's apprentice in 1644. He had two older sisters, Anne and Katherine.

▶ **Continue ahead along Hooke Hill to the roundabout by the Co-op.**

Here there is a plaque to Robert Hooke.

Hooke Stone

▶ **Turn left along the pavement past the Co-op and the End of the Line Café is a few paces further along adjacent to the Garden Centre. Walk a little further along the road and then cross the road to a grassy opening about 20 metres past West Wight Printers. Turn right and take the right-hand path along the stream to a wooden bridge. Cross the bridge and continue ahead to the road. Cross over to the grass verge, turn right and go ahead to a footpath on your left. This area is called Afton Marsh. (In very wet weather there is an alternative path further along which bypasses the Marsh).**

Many of Hooke's ideas and observations were presented in his first major publication,

Micrographia, a summary of most of the scientific work he achieved before the age of thirty. In the opinion of R.S.Westfall, the American historian of science, *"Micrographia remains one of the masterpieces of seventeenth century science"*. With observations from the mineral, animal and vegetable kingdoms, the book demonstrated what the microscope could do for biological science. The book describes how Hooke developed the microscopes he used and contains illustrations of his observations with the instruments. Some of these, like those of the snowflake and the flea, are both magnificently detailed and beautiful. Many plants and animals that are illustrated in Micrographia can be found along the Afton Marsh stretch of the Trail, among which are: common nettles, bees, spiders, moths, male and female gnats, tree ants, moss, deciduous foliage, water mites, poppy seeds and silver fish.

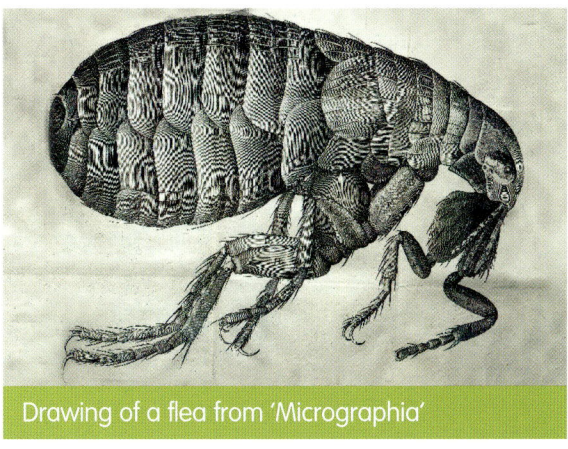

Drawing of a flea from 'Micrographia'

At the end of Afton Marsh, turn left (signposted "Footpath to Freshwater Bay") and continue ahead, taking the left-hand fork at the T-junction to the gravel road. Continue ahead, past the Sandpipers Hotel on your left, to the main road. After passing some former Coastguard cottages on your right, cross the road and turn left for a short distance towards the shelter and views of Freshwater Bay.

Freshwater Bay

Crossing the road towards the shelter, a glance to the right reveals Fort Redoubt sitting on the headland above the Bay. It was built in 1855 to stop the French landing in the Bay, in order to protect the Solent from land invasion. The original fort consisted of a ditch behind the headland. The flat area visible below the current buildings formed the parade ground. Beneath these are carved tunnels and store rooms, safe from mortar shells.

Freshwater Bay

The western end of the Isle of Wight is rich in fascinating geological formations containing the bones of dinosaurs. Hooke probably dug fossils along the nearby beaches of Totland and Freshwater Bay, and from the chalk cliffs that run from Compton Bay to the Needles. Watching the cliffs of the West Wight crumbling into the sea led the adult Hooke to an idea of fundamental importance, *"that a great Part of the Surface of the Earth has been since the Creation transform'd, and made of another Nature: that is many Parts which have been Sea are now Land, and others that have been Land are now Sea; many of the Mountains have been Vales, and the Vales Mountains"*. He described observing on the Isle of Wight a layer of sea sand extending above the water level to a height of sixty feet, in which was embedded oyster shells, limpets and periwinkles. These explorations may well have planted the seeds of his later ideas about the extinction of species, the formation of mountains and dramatic changes in sea level. The changes in the cliffs may also have led him subsequently to doubt the presumption of the time that the world was unchanged from when God created it six thousand years earlier. These were bold ideas in the Christian views of the seventeenth century.

Hooke maintained contact from London with members of Island Society, like the Deputy Governor, Sir Edward Worsley. He returned to visit the Island on the death of his mother in 1665, and probably used this visit to look for fossils. Hooke's lectures on earthquakes show him to have been among the first geologists and a pioneer in the field. Many of his ideas were not only overlooked in his day but were plagiarised by others.

Retrace steps to the road by the Albion Hotel, and continue ahead past some public toilets and then past Dimbola Lodge to the thatched church of St Agnes.

St Agnes's Church

St Agnes's Church

St Agnes's Church was built in 1908 on land given by Hallam, Lord Tennyson, the elder son of the famous poet. The stone walls for this building reputedly were taken from the old cottage that had been Robert Hooke's home.

▶ **Take the left hand road uphill (Bedbury Lane) at Orchards corner shop. Continue ahead to a footpath on your right just past the entrance to the Farringford Hotel. Turn right through a kissing-gate and continue ahead along a tarmac path (F41) to Pound Green. Cross the road onto a path through a triangle of grass, and continue along Queens road to the T-junction at Brookside Road. Turn right along a tarmac path, with a stream on your left, and through the car park to Moa Place.**

Kissing Gate

To the left of the car park is the **Ye Old Village Clock Shop**, dealing in antique clocks - This shop has no direct connection with Robert Hooke but nevertheless illustrates the way his early interests developed. His notebook describes how, as a boy: *"Seeing an old Brass Clock taken to pieces"*, he attempted to imitate it and made a wooden one that would go. This developed into a life-long interest in clocks, leading to two of his most important inventions:

Balance-Spring Watch

It was the Dutch scientist, Christian Huyghens, who made the first pendulum clock in 1657. Hooke wanted to improve the accuracy and reliability of such clocks so that they could be used to help navigation at sea. Before 1660 he had applied a spiral spring to the balance wheel of a clock so that the spring's natural oscillation would serve as a regulator of the mechanism, and had improved the anchor escapement. He demonstrated a spring-regulated watch to Lord Brouncker and others but, although advised to patent the invention, declined to do so. During the 1670s Hooke collaborated with Thomas Tompion, the father of English watchmaking.

In the course of this work, he made the discovery for which he is most well-known today and which many of us may remember from our school-days, Hooke's Law of Springs. He stated this as *'the power of a spring is proportional to its extension'*, by which he meant that a spring is stretched in proportion to the force or weight pulling it.

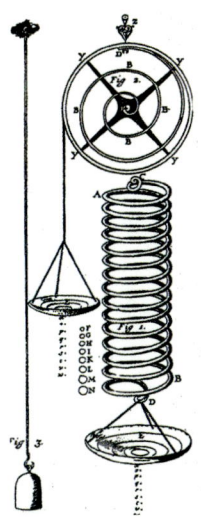

Hooke's Law

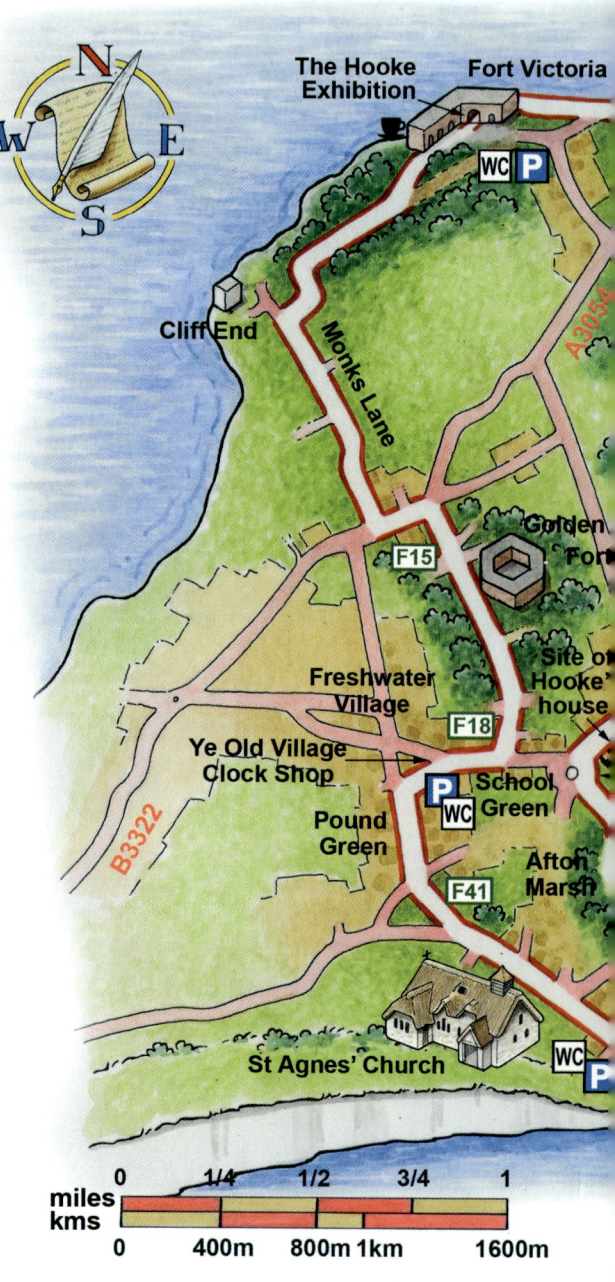

The Robert Hooke Trail

Universal Joint

In 1677 Hooke *"contrived motion of the pendule without noyse by universal joynt"*. This device, which he used to allow a pendulum to move noiselessly, is today used in the drive-shaft of a car for linking the engine with the wheels, and also in the supporting blocks for an oil rig, where the joint can be as large as a house.

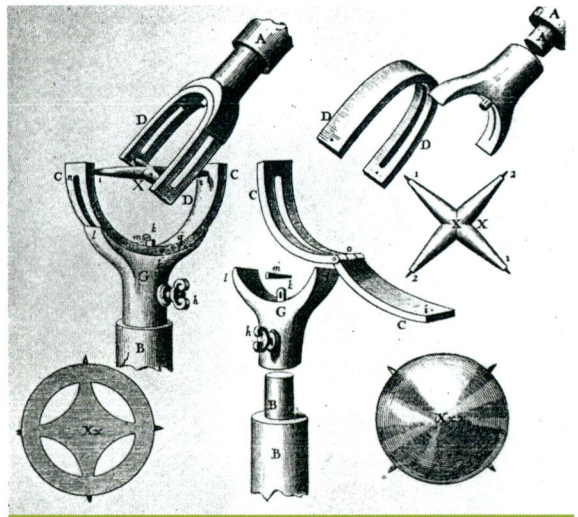

Universal Joint ("Hooke's Joint")

▶ **Exit the car park by the toilets and turn right along the pavement with the stream on your right, past the library to a pedestrian crossing. Cross the road and turn right towards All Saints' School. At the school entrance turn left onto a tarmac path (F18) and continue uphill to the path cross-roads. Continue (sharp right and sharp left) on a dirt track to Golden Hill fort. As you near the top of the rise take a few minutes to enter the gravel track on your right for 20 metres, and admire the views of the downs farmland to the east of the Island.**

Golden Hill Fort was completed in 1868 as part of a string of forts known as Palmerston's Follies.

The Robert Hooke Trail

Golden Hill Fort (hexagonal building)
Photograph by Kevin Clarke

▶ **At Golden Hill car park continue ahead (F15) through the industrial complex to the main road. Cross the main road and turn left down-hill to the T-junction with Monks Lane. Turn right into Monks Lane and continue ahead, past the entrance to the Brambles Holiday Village, and uphill to the entrance to Cliff End. Turn right (signposted "Coastal Path") onto a path between two fences and continue ahead around Cliff End to a gravel track and a View Point.**

Here, there is a view across the Solent to Hurst Castle. Built during the reign of Henry VIII, some of this Castle would have been in existence in Hooke's time.

▶ **Continue ahead downhill and turn right along an old roadway, ignoring the path to the shore on your left, and continue ahead along the gravel track to a path T-junction. Turn left (signposted "Refreshments and Toilets") and downhill to the back of Fort Victoria.**

Island Planetarium, Fort Victoria

This is the site of the Robert Hooke exhibition. Hooke made many observations of sunspots, spots on Jupiter and the moons of Jupiter. He also made a map of the surface of the moon which agrees well with modern charts. He gave a lot of thought to the questions raised by comets and their orbits. This may be partly the reason he proposed an inverse square law for gravitation. However, solving this problem mathematically could not be done without calculus and the laws of motion developed by Isaac Newton.

Fort Victoria

▶ **This completes the Trail, which we hope you have enjoyed. There are other areas on the Island that have some connection to Robert Hooke and which are worth visiting in their own right.**

Yarmouth Castle - This is a possible detour near the start of the Trail. Tucked away down a passage by the Wightlink ferry, stands Yarmouth Castle some of which stood in Hooke's time. As a result of the French sailing up the Solent in 1545, Henry VIII ordered the building of the Castle as a coastal defence. It is now in excellent repair, surrounded by the sea on two sides, and a moat on the land side.

Alum Bay has been described as having on one side a wall of glowing chalk and the other a barrier of rainbows. This gentle coast is composed of the famous Alum Bay coloured sands, 20 or more different shades showing up the brightest after rain. Hooke may well have collected coloured sand from the stratified cliffs here.

St. Catherine's Lighthouse - In 1663, Hooke read a paper to the Royal Society on the observations necessary for *"making a history of the weather."* He followed this up by constructing a wheel barometer and a spirit of wine sealed thermometer (whose zero was set at the point where pure water began to freeze). His observation that a beard of wild oat curled in dry weather and straightened when it was humid also gave him the idea of making a hygrometer to indicate humidity and dryness of the atmosphere. He later devised an anemometer to measure wind strength. With these instruments, he began keeping some of the earliest meteorological records. Hooke's role as one of the founders of the science of meteorology was commemorated when the Hooke Institute for Atmospheric Research was set up at the University of Oxford in the 1980's.

St. Catherine's is the embodiment of Hooke's vision. In its grounds, the lighthouse has an automatic weather station reporting every hour to the Meteorological Office, and is the representative station of the Wight area.

Minster Church of Sts. Thomas, Newport - The Church is situated in a square that was once the corn and meat market. The square also contains God's Providence House with a fine 18th century porch. Now a restaurant, it stands on the site of an older house where the Plague came to an end in 1584. Robert's elder brother John, who had a grocery business in the High Street, was twice Mayor of Newport before committing suicide in 1678, perhaps because of debts. John's name appears on the list of Mayors displayed in the Church. At his death Robert Hooke owned land on the Island. The County Archive has a mortgage agreement dated 1684/1685 between the Town of

Newport and Robert Hooke, covering a small number of pastures where the multi-screen cinema and car park are now.

Hooke's Scientific Career

His adolescence in London

In October 1648, when Hooke was thirteen, his father died of *"a Cough, a Palsy, Jaundice and Dropsy"*. This was a time of unusual excitement on the Island due to the imprisonment of King Charles I in Carisbrooke Castle after his defeat in the Civil War the previous year. He was executed in London two months later on 1st December. By this time, Hooke, too, was in London. Being by the standards of the time old enough to make his own way, he had taken his inheritance, £40 from his father and £10 from his grandmother, and set off for the mainland. Initially, Hooke intended to take up an apprenticeship in Westminster with the great Dutch portraitist, Peter Lely, perhaps with introductions from a well-placed friend, but the position did not suit him, and he stayed only a few months.

Robert Hooke and Dr Busby

Instead, Hooke moved to Westminster School, living from 1649 to 1653 with its severe headmaster Dr Richard Busby. The school specialized in producing clergymen and, to quote Stephen Inwood, Busby *"left the mark of his rod on the backsides of at least sixteen future bishops"*.

Although Hooke was not often seen around the school, his fellow pupils included the future poet and playwright John Dryden and the philosopher John Locke. At the school Hooke mastered mathematics, Latin and Greek, as well as studying other languages and trying to invent ways of flying.

At Westminster, he began to develop a deformity, a curvature of the spine. This problem, which may have had its origins in genetic or nutritional factors, grew worse as he got older. In his middle and later years, he was described as having a thin and crooked body, an over-large head, sharp facial features and protruding eyes.

Life at Oxford

From Westminster School, Hooke went on to Christ Church, Oxford, where, as his money had run out, he had to pay his way by acting as a chorister and as a servitor to a richer student. This was not an unusual way for poorer scholars to make ends meet: Isaac Newton paid for his studies at Cambridge in the same way.

At Oxford, Hooke learned mathematics and mechanics from Dr. John Wilkins (who was also Oliver Cromwell's brother-in-law) and dissection from the physician, Dr Thomas Willis. It was Willis who recommended Hooke to the wealthy amateur scientist, Robert Boyle. His appointment as Boyle's assistant was to set the course of Hooke's career. From 1656 to 1663 Hooke aided Boyle in his famous work on the pressure of gases.

Christ Church, Oxford

His Life's Work Begins

In 1662, Hooke was appointed Curator of Experiments of the Royal Society (initially unpaid); just at the time when a new practical approach to scientific investigation was developing. Each week he had the task of devising one or more experiments to be shown to the Fellows of the Society, most of whom were rich men with an interest in science. During his forty years in this post, Hooke devised an astonishing diversity of ingeniously simple experiments which led to many of the scientific advances of the seventeenth century: he may have inspired some of Isaac Newton's later discoveries and devised the means by which Christopher Wren could build the dome of St. Paul's Cathedral, as well as designing mechanical systems for the great clockmaker Thomas Tompion. In 1664, he was also appointed Professor of Geometry at Gresham College, a position he held for the rest of his life.

The Air Pump

Robert Boyle wanted to make an air pump, which was easy to operate together with a vacuum chamber in which experimental devices could be placed and observed. This presented great practical difficulties, especially in preventing leaks from around the piston and from the sealed lid of the vessel, problems others had not been able to overcome. Hooke *"contrived and perfected the air pump for Mr Boyle"*. However, he has often been accused of claiming credit for almost anything, and so this might be disbelieved, but Boyle, unusually for his time, openly acknowledged that his assistant had made him a pump that worked. Using this apparatus, which could create high as well as low pressures, Boyle and his assistants carried out many experiments on the qualities of air. This enabled Boyle to establish the famous law that bears his name:

"at constant temperature the volume of a gas is inversely proportional to the pressure exerted on it".

Hooke and Boyle with the Air Pump

Fortune from the Fire of London

Hooke was much more than a great scientist; he was also the architect of some of London's greatest public buildings. Following the Great Fire of London in 1666, Hooke was made City Surveyor under Christopher Wren and, in this position, was instrumental in restoring the City. He played a major role in the design and building of the Monument. Hooke re-built many of London's churches at this time but unfortunately only three of these remain standing - St. Benet Paul's Wharf, St. Martin Ludgate and St. Edmund the King.

All this work made him very wealthy. His salary from the Royal Society and Gresham College, when it was paid, was £80 a year, but after his death a great iron chest in his rooms was found to contain £8000 in money and another £300 in gold and silver. This is almost £1 million at today's values.

The Great Fire of London

Over two centuries of Obscurity

Although he never married (his Professorship also required him to be single) and his character was sometimes difficult, Hooke was very sociable, daily meeting friends and associates in City taverns and the new coffee houses. So, with all these achievements, why is he not famous today?

Robert Hooke had two misfortunes. Firstly he lived at the same time as Sir Isaac Newton, a scientist who reduced the motions of the moon and planets to order, and in the process developed a whole new branch of mathematics, the calculus. Newton also carried out fundamental experiments on light and invented the reflecting telescope. Sir Isaac Newton was head and shoulders above almost any scientist who has ever lived. Hooke's second misfortune was crossing Newton, an easy